D0984875

$18.00
(signed!)

Front Cover - Winter in Swaledale
Back Cover - Duncan and Tony

HILL SHEPHERD

HILL SHEPHERD

A Photographic Essay by

John and Eliza Forder

John & Eliza Forder

ACKNOWLEDGEMENTS

We would like to express our gratitude to Colin Baker, Managing Director of Frank Peters, for the help he has given us while making this book. It is through his concern for recording life in Lakeland and the Dales that the idea for HILL SHEPHERD was conceived, and it is with his support that it has been realized.

We would also like to acknowledge Ronnie Mullin and Steve Moran at Ram Design for all the care and trouble they have taken in the design work.

Above all we want to thank the hill farmers, who gave their unstinted help, encouragement and hospitality throughout the year we photographed. HILL SHEPHERD is dedicated to them all.

First Edition September 1989
Second Edition January 1990

ISBN 0948511 65 6

Published by
Frank Peters Publishing Limited.
Kendal, Cumbria

Printed by
Frank Peters Printers Limited.
Kendal, Cumbria

Designed by
Ram Design Associates Limited.
Kendal, Cumbria

CONTENTS

THE PHOTOGRAPHS

INTRODUCTION

It would be foolish to pretend that observing and photographing an upland community at work could generate a full understanding of a life-style that stretches back over centuries. All we can do is pull aside a veil or two for glimpses of a life that has preserved its privacy from the outside world, in part because the terrain makes for isolation, but also because the people maintain a natural wariness towards intruders. Where print and picture stop, a sensitive imagination must take over.

How can we appreciate the frustration felt when mist is down day upon day and sheep need gathering, when rain teems and the wool needs to be dry for clipping? "If ye put 'em in t' barn for long, they get all mucked up." How can we understand the self-reproach when too long a cat-nap at three in the morning means that a good calf is born dead because we were a half-hour too late in attending the heifer? And "that bloody shear bolt. Ye'll 'ave made thirty, forty bales with weather threatenin' and t' baler breaks down; ye put in t' new bolt, get two more bales and bang, crash, kkkhh . . . ruddy thing's gone again." Blistered hands, the arthritic knee, the fusion of dip smells, muck smells, silage and wet wool go hand in hand with the vibrant air of the fell, the pride felt at a good lamb sale and the satisfaction of hot, sweet tea and 'bait' after dipping a few hundred sheep.

The watching of the weather, and planning accordingly, of coping with personal mishap and the apparently irrational changes in government policy, create a degree of uncertainty that is met with an acceptance of life, death and a steady pace of change among hills that seem passive and changeless. The uncertainties and the unpleasantnesses give the colour and provide the spice. They are not concealed and cannot be by-passed. As a young, pretty shepherdess said,
"Weather is life, in't it?"

The image of the shepherd striding the fells with dog and stick is but a fractional aspect of a life that is diverse and unpredictable. He must turn vet, mechanic, waller and form-filler as each situation dictates. Varying ways are handed down from generation to generation but are modified to suit the times. Father will use hand clippers as that's what he's used to, while son will use electric as that's how he's learned. From farm to farm and dale to dale methods differ and old habits die hard. The intricate pattern of modern practices reflects a past that is rich in tradition.

The tractor has brought one of the biggest changes in past years. The horse and sled were finally abandoned in the fifties, while the advent of machinery has enabled siloing to get well established in the eighties. Among hills where, owing to wet summers, hay must at times be made as late as October, big bag silage is now accepted as the safe way of ensuring winter fodder. Hay bales may well become a feature of the past, just as the trial bike is rapidly becoming a feature of the present. When the inquisitive walker, escaping from his urban dwelling, complains that his peace is being disturbed, the retort is straight.
"If ye want to carry poorly sheep down t' fell yesel', ye're welcome. Mesel' I use t' bike."

The fell shepherd is lithe and energetic, practical and adaptable. He is caring and he is proud and will enjoy showing off a good ewe or a fine tup. (It's 'tip' in Cumbria and 'tup' in Yorkshire). Not easily discouraged, he will persist where others would give up. You know where you stand with him - or you think you do! He has a droll sense of humour and tales are told . . . sometimes against each other.
"Ye need a broad back and a smile to be a fell shepherd!"

GATHERING SHEEP AND TIP TIME

"I allus think our year starts in November, when we gather sheep in at tip time," says a Lakeland shepherd. In one of the five main gathers during the year, the sheep are brought in for a "dose and a change of pasture" before being put to the tup.

Sheep like to be out on the fell, especially in summer, and to gather them in is an art passed on from one generation of shepherds to the next. Nowadays father and son may work a farm that used to keep three or four full-time shepherds, or an extended family, but the all-terrain bike serves as a valuable replacement for the hands that used to help. This can successfully be used for gathers on lower fell farms where land is gentler, but on the high Lakeland fells the terrain is more hazardous and the shepherd has only one way to get about, and that is on foot. There are crags, scree and wide open fell to contend with, and the skill of the shepherd and his dogs should not be under-estimated.

There are no footpaths and no shortcuts. His crook is used to steady himself on the steep fellside while attention remains focused on the movements of his dogs and sheep. Underfoot the ground varies from rocks and stones to bog and slippery scree; yet shepherds can move fast and they need to. Despite the steep incline up the fell and the rough ground, speed and timing are important, for otherwise the sheep will soon scatter. Weather must be watched too. A morning can be whiled away waiting for the sky to become a lighter shade of grey if the gather is near to home, but for day-long gathers a decision about departure must be reached the previous night in readiness for an early start next morning. The heavens can throw everything at you on the fell. It is not unknown for snow, hail, rain, mist and sun all to come and go within the space of a few hours.

The July gather for clipping time is the hardest. The ewes are weighed down with wool which makes them difficult to stir if the weather is hot, "and nine times out of ten it is hot when ye want to ga to t' fell." The lambs are also with them, which doubles the number of sheep that need gathering, so the shepherd takes all the dogs he has. "And they are dogs too – not them fancy dogs that peek and peep abut. How can a dog work among brackens on its belly? You need summat with a bit o' bark."

'Cur' or 'barking' dogs have been used in the Lake District for generations and are bred especially for the job of shifting sheep in difficult terrain. They are

Blue-faced Leicester Tups

trained to bring sheep away from crags, cliffs and undergrowth, while a sheepdog from lower, 'cleaner' country may resort to sinking its teeth into them through sheer frustration. Herdwick sheep, local to Lakeland, will outwit shepherds and dogs if they can.

As many as four or five men are needed for mass gathers on common fell land. Each covers a different part of the fell, and though they work as a team, they may use different methods for running their dogs. One may prefer to let them go as a pack, while another may decide to let three work on the way up the fell, and keep two coupled back in reserve.
"Then ye've two fresh'uns to flog sheep back hyaam, and they seem to give more bounce to them that are faggin' a bit".

Inevitably there are losses. Some sheep will be found dead at the bottom of high crags, while others may escape with a broken leg. It is not uncommon to have to carry an injured sheep off the fell. It is a time to watch for stock that are ailing, stuck on ledges and in bogs. In a big gather a few are bound to get missed and this entails a return journey to the fell to "rake around for the odd'uns." Even so, Herdwicks can be found with wool on their backs way past clipping time, having escaped being brought down several times.

The climb up the fell begins. With four dogs running and a young one to train, the shepherd steadily negotiates his way along a narrow track keeping a keen eye out for lone sheep that may be lurking among brackens or camouflaged against the grey scree, for they must be driven across to join the others. It is a stiff pull up. The occasional detour is made to check nearby gills until sufficient height is reached to begin traversing the tops. Now speed is gained and, sure-footed, the shepherd skirts around the outside, shoving sheep across to the next man and to the next and so on, until it is possible to close in on them, heap them up, and turn them back for home.

On craggier fell the shepherd must scour the land from the tops, one crevice at a time, driving sheep out of the gills and down the uneven slopes as quickly as he can. A few sheep here and a cluster there slowly start picking their way through the scree towards the brackens where they disappear from sight. The shepherd, now waist high in the undergrowth himself, whistles for his dogs, and between them they create enough commotion to chivvy the sheep further on. The dogs leap high as they plummet through. Small groups of sheep now

appear from different directions, some still a fair distance up the valley, following on in little lines. Lambs dart about nervously, working the dogs twice as hard and trying the shepherd's patience.
"Once a visitor thowt me dog's *name* was Bastard!"

The shepherds soon sense it is time to close in, and start zig-zagging back and forth encouraging the singles to join the rest. At last it all begins to come together. The sheep congregate in one enormous mass near the fell gates knowing it is time for home.
"Ga hyaam, ga hyaam," echoes the voice of the shepherd.
Two gates, a packhorse bridge, and a narrow lane still to be negotiated, and several hundred sheep are safely brought back to the farmstead, where they are coaxed into the pens to await sorting. There is a feeling of quiet satisfaction as another gather is done.

Each time sheep are brought in, they are put through the 'shedder', a network of swing gates and fences, set up so that sheep can be divided into separate lots. If the land is unfenced, there may be a fair few strays and these will be taken out first. In past times shepherds would gather at 'meets' where the strays could be returned to their owners, and these occasions were treated as festive gatherings. Now all that is needed is a phone call and a journey in the pick-up. Before tip time the ewes are checked over to make sure they are in good fettle: if their udders are unsound they will be unable to suckle lambs and if teeth are defective they will not find it easy to survive the winter and spring. Older ewes are drafted out, for stock must be kept young on these hills. These 'draft' ewes of four to six years old will move on to an easier climate and continue their breeding cycle there, for they still have "a gay bit o' youthful life in 'em yet. They follow water, downhill . . . where we all have to ga sooner or later."

The choice of a tup is a critical decision.
"When I choose a tup, I look for yan that's garn t' get rid o' faults in me yows."
Not only experience but also instinct plays a part. The shepherd will examine a tup thoroughly, but then uses a sixth sense to select the right one for his sheep. When a tup has proved himself and sired good lambs, the shepherd will continue to use him for as long as there is no danger of his crossing with his own offspring. On occasion, if a tup has a fine reputation, neighbouring farmers will pay for the privilege of bringing their best ewes to him. A fine Swaledale tup from good stock may fetch several thousand pounds in the auction ring and

the October ram sales, prior to tupping time, are a draw to shepherds from all over England. There are many factors that are taken into account: stance, colouring, mouth, wool condition, and most important of all – pedigree.

The pure-bred sheep, being hardy, are kept on the fell but on more marginal land, sheep are also crossed between one breed and another. For instance, when a Blue-faced Leicester tup is put to a Swaledale ewe it produces a mule, which is popular on southern farms because it is good for breeding fat lambs. Teeswaters, Suffolks and some continental breeds are all crossed with the Swaledale to produce lambs with varying characteristics.

A tried tup will have as many as sixty ewes or more to run with. He will be given a good feed beforehand to "fitten him for t' job", and a rest at the end to help him to recuperate, for it is no light work! Dye or 'ruddle' is painted on his underside so that when he mounts the ewe a vivid patch of colour appears on her back, acting as an indicator as to which week she will lamb and also to confirm the tup is working.

Tupping is done in the lower pasture or intakes, enabling the shepherd to visit his flock and ruddle the tup, take ewes out that have already been marked or simply gather up the ewes around the tup to remind them all of their task. As an older shepherd explained, a ewe can lamb anywhere, but it cannot get tupped anywhere, so wind or hail or Christmas day he would make sure that his ewes were not straying far from the tup. But not all the ewes will take to the tup within the first few weeks so when they are turned back to the fell 'jacks' or 'chasers' are sent with them. These are tups that may be slightly inferior but "they'll ga abut like young fellas here an' there an' everywhere, acting like as a sweeper."

Swaledale ewes are put to the tup as shearlings (two year olds), but Herdwicks are usually held back for another year so as not to risk stunting their growth. They cannot mature as fast on the higher fells. So these shearlings, or 'twinters' as they are called in Lakeland, have to be kept away from the tups that are running with the ewes. If the farm is not well fenced the practice of 'clouting' may still be used.
"Ye get a lile bit o' clout, old sacks an' sek like, cut it into bits rather bigger than a postage stamp. One fella sits on t' stool and holds twinter, while t' other sews bit o' clout o' er its lady parts with tail clapped down agin it an' all, and that clout stops in spot 'til March. Double protection as ye might say!"

There are, however, no rules and no two shepherds work alike. Each will respect another's methods if they bring good results. One might clip the ewes' tails on either side to help tup along with his job, while another may prefer to let tails be, claiming that his tups manage anyway. Another shepherd may put a few hundred sheep and a number of tups together, and not bother with ruddling or gathering them up.
"I used to gather them to tips but tips would fight and carry on. I look at it that if a yow's for a tippin' and a tip comes up and can't tip it, another yan'll come up and tip it. Leave 'em alone, that's what I say. Let nature take its course!"

Indeed when there may be as many as two thousand sheep to watch on some of the higher Cumbrian farms, the methods are bound to differ from those used on smaller Dales farms, where there are fewer sheep, but richer pasture enables dairy cows and suckler calves to be kept for additional income.

Tupping time allows for a little experimentation, trying a new tup here or a different cross there; but above all it is a time to keep tups and ewes in good health, so that there are as few 'geld' (barren) ewes as possible at lambing time. Except for the higher farms, the sheep will be returned to the fell by the middle of December to weather the winter as best they may. The shepherd can now turn his attention from sheep to turkeys and other Christmas fare.

WINTER

There is little romance about winter on the fells. Paintings depicting the gentle snow scene, the stooped shepherd with the hay bale, or sheep sheltering at twilight have little to do with the reality of facing the shear bleakness and desolation of winter days on the tops. Leaden skies, mists, gales and sleety rain may be more common than snow and the picturesque.

There are the notorious winters when drifting snow blows in time and again and digging a way through it is like digging quicksand - it fills straight in again. But these winters do not come frequently. There are also the exceptionally mild winters when more cold weather would be appreciated, for there is a belief that "frosts kill off the bugs". However it is more likely that the months from January to April will bring the full range of weather in no logical sequence. New Year's Day can feel like the first day of spring and April can bring the bitterest of winds and heaviest snow falls of the year.

Not only is the weather unreliable with regards to timing but it also alters considerably as height is gained. The valley bottoms will have an entirely different experience of winter from that of the fell two thousand feet higher. The wind that hums through the trees around the farmstead may become a ferocious gale on the tops, forcing the shepherd to bend double to withstand it. The rain falling gently on the slated roofs drives its way up the gill to become sleet or hail, and then slashes fiercely across the face. Thin drizzle in the village concentrates itself higher up to become thick mist and each rock, hummock and ridge appears larger than life like ghostly images, making it easy to become disorientated despite familiarity with the terrain. A smattering of snow becomes a ten foot drift.

It is this unpredictable nature of winter that makes shepherding difficult. The fell shepherd must be ever-vigilant, watching the weather as he watches over the sheep. His lamb crop will be dependent upon how well his ewes survive the winter and he cannot afford to take risks. A weather forecast, although a valuable guide, may not necessarily be reliable as to either severity or timing. One area of fell might get a bad snow storm, while a couple of miles across the ridge there may be scarcely a covering. A shepherd who likes to bring his sheep down to the bottoms in case of snow must be the judge and decide for himself when to fetch them in, and "it's an easy thing to miss."

Despite the hardiness of fell sheep (a ewe may lose twenty per cent of its body weight and still produce a lamb), their feed in winter needs supplementing. If there is to be an easy lambing and plenty of ewe milk, extra hay, cake or feed-nuts must be available for the sheep. Fell land varies in quality. When 'the mosses' come through early in the year, they provide extra protein and older shepherds have claimed them to be as good as cake - though few today would take that gamble. Heather or ling fell is rich in nutrients and sheep thrive well on it; they can also knock snow off it more easily, enabling them to feed in bad weather, for it is energy the sheep are after. From farm to farm and fell to fell, methods of supplementing diet will vary according to the needs of the stock and the quality of land. It will be rare to find two shepherds that feed their sheep alike.

In the northern Pennines it is not uncommon for snow to cover the moors for three months or more at a stretch, and the daily round of filling the hay racks can begin in December. The snow plough and snowmobile have facilitated this task, enabling the shepherd to move about more quickly and efficiently in bad conditions to feed and check on his sheep. As these vehicles help with the task of winter feeding in the Pennines, the helicopter is used in the Lake District to drop feedblocks at certain places on the high fell at the end of November. While some farms make use of this assistance, others will leave fodder further down and the gates open so that, as the weather worsens, the sheep are lured away from the fell.

Herdwicks are the hardiest of sheep and, though some may choose to retreat, others will brazen it out up the fell for most of the winter. When the storm comes they turn their backs to it and shelter wherever they can. Shearlings must be watched too, for it takes a while for them to sense the right time to come down, and they have not yet learned that food awaits them.

Although the fell is higher in the Lake District, farms, for the most part, nestle in the valley bottoms where protection is afforded. In the Pennines, however, some farmsteads are situated in open country well over a thousand feet up, and here the winters can be awesome. It is the wind and snow together that cause the major problems. The shepherd will try to bring his sheep nearer the farm buildings in the event of snow, so that he can tell more accurately where they will shelter, but there are no guarantees.

"After yan snow blow, I took dog wi'me and I prodded in all t' usual spots but there wasn't a sign of t' sheep anywhere. After a lile bit, dog got to scratchin' right in t' middle of t' field where there was a bit of a dip. Sheep'd never gone there before but some dogs have t' knack o' sniffin'em out. I began digging and pulled two lots out, but when melt came, I'd missed another lot below, and they were all dead. Ye allus have to watch for that. Snow can shift all t' time, covers walls an' owt like that - it can all look so terrible different. It's a hell of a different story than farming on an inside spot. On t' fell it's hard to know where to start."

A sheep on its own may survive cocooned in the snow for as long as two weeks. The trouble comes when either they get blown together or they huddle near each other for protection: they can then trample on one another and suffocate. Those that survive emerge one by one from their little igloos, with bare patches of fleece on their backs where, during their days below the snow, they have pulled at their wool for extra nourishment.

But snow is not the only problem. Interminable rain may be worse, for sheep do not like wet weather; they prefer it cold and dry. A ewe will avoid lying on wet, muddy ground for as long as possible, but eventually her legs will tire and she will be forced to rest. If she gets chilled and distressed, there is a danger that her lamb will be born weakly or even dead.

As winter draws on and lambing time approaches, supplement is stepped up. Fodder does not come cheap and now there are some shepherds who are making use of the facility of having their ewes scanned to discover which ones are carrying twins. Feed can then be apportioned correctly, giving those sheep with twins more, singles less, and leaving the geld sheep to fend for themselves. Hill farming means taking good care of one's resources.
"After all, we're here to make money."

The winter months are not just spent feeding and over-seeing the sheep. It is a time to catch up on jobs around the farm. A barn wall may need rebuilding, extra sheep pens putting up and so on. Hedges must be laid before the sap rises and there are always plenty of walls to be repaired. When the weather is kinder, many a day can be spent on the fell building up the gaps and replacing top-stones. With cows milked and breakfast done, the shepherd will set off to the crag taking drink, bait and the dogs for company, for he will not return until the light fades. It can be a lonely job, but there is pride in the finished work. Fencing is quicker but costly and does not last as long as a dry-stone wall.

The cattle have been inside since November, for by then there is little for them to eat in the pasture. They spend the winter either in hill barns or in large modern sheds that are built near the farmstead, which makes it easier to feed them silage. Their muck is good, valuable stuff and from December will regularly be spread on the lower fields to encourage the spring growth.

The shorter days mean that more time is spent inside, but there are always forms to fill in and the income tax. Yet although there are plenty of tasks at hand, there is less sense of urgency. It is a time for fireside planning, the rum that is left over from Christmas, and a bit of dominoes and darts. Ice patches still lurk on the fell roads and the wind gusts unexpectedly; folk are not bothered about venturing out far.

"It's not oft I get entertained at this time o' year. Not so many folk about, it's just me an' t' old bride, that's all."

The only sure way of telling spring is on its way is by the lengthening of the days. The emergence of daffodils means nothing: they, unlike shepherds, can be fooled.

LAMBING

Dalesbred

March 20, April 1, April 26: specific dates loom large in the shepherd's calendar at the onset of his busiest and most trying time of the year. It is now that his skill and ingenuity are tested to the full. The higher the farm, the later the date, for tupping time is delayed accordingly to allow a chance for the arrival of more settled weather before the first lamb comes. The cross-bred lambs are used as the year's cash crop and it is usually planned for these to arrive earlier than the pure-breds, in order to give them a chance to reach a good size before the Autumn sales. It will be the end of May before the last black Herdwick lambs appear.

During the lambing weeks there is no day off and little rest. The shepherd's mind is always with his ewes and lambs. There is the shearling, lambing for the first time, shifting restlessly by the wall, scratching at the ground and moving from side to side. She may need tending in the next hour or so. Another shearling needs tying up, as she butted her lamb down the field in disgust when it sought her teat; leaving her penned up with "the scrap little thing" should make her more accepting of it. Another ewe got away with the first lamb and abandoned its sickly twin, which adds one more to the number that already need bottle-feeding; and so it goes on, week in, week out.

As soon as it is light, the shepherd pulls on his over-trousers and worn jacket that now carry that distinctive smell of lamb and milk, takes up his lambing tackle and, without bothering to wait for a drink, hurries out on the first round of the day. With a steady but quick pace he goes to the top field where there are some first time lambers that may need supervision. One has lambed in the night and already the little one has strength enough to provide a chase as the shepherd tries to catch it with his crook in order to check it over. It seems healthy and is clearly sucking well so he lets it go back with its mother.

Otherwise all seems quiet, so it is on to the next lot, where the shepherd has seen from a distance that a ewe is in trouble. She is down and has clearly been struggling for some time, so despite the rule of always giving a sheep plenty of time, the shepherd decides it is right to intervene. Among his lambing tackle he carries lambing oils enabling him to examine the ewe internally with least fear of infection or letting his nails tear at the 'lamb-bed' or womb. Gently and easily he slips his hand inside and discovers that the two forelegs are bent back. Very carefully he twists the lamb round so he can manoeuvre the legs forward below the head and, once this has been accomplished, a wet bundle emerges into the cool morning air. There are clear signs of relief from the mother as she licks the yellow mucous membranes from the lamb's face, enabling it to take its first breath. The licking also stimulates the circulation of the rather pathetic object that already is struggling to its feet in search of teat and comfort. The shepherd checks that the ewe has milk and, satisfied that there is no more for him to do, moves on, for there are still three more fields to look over.

There are no more immediate problems, only a twin that needs a top-up with the bottle, and certain ewes are noted as the ones to watch on the next round. A light drizzle is turning into rain, but the first tour of the day is done and it is back to the kitchen for a mug of tea and porridge. As he reaches for the pot on the stove, a sniffle at his feet reminds him of the half-starved creature he rescued the night before; warmth seems to have revived it.

A vigilant shepherd will try to get round his flock four or five times a day, and if there is a specific problem it might entail his making an extra visit at night with the flashlight. The use of a bike will save time and his legs, for the other farm jobs cannot be abandoned. The cows are to be milked and mucked out, eggs collected and washed, the early lambers taken back to the fell and the later ones brought down. Every lamb and ewe that dies is a financial loss to the shepherd, so he is well aware that he is nurturing the year's income as he nurtures his flock.

Twins are a bonus - but not too many of them, especially among the fell sheep. "All ye end up with at back-end is two, lile, wee, scrappin', pot-bellied bloody things that nay bugger wants. It bothers them to trail abut never mind owt else. Ya good'un, that's what we want, that can get away back t' fell with its mother." Twins are better appreciated among the cross-bred lambs, which need to be kept on inside land to be sold later on in the year; the pure-bred sheep must return to the fell quickly, with strong healthy lambs fit enough to cope with the terrain.

April is a time of transition. The curlews have returned to argue over territory before their courtship rituals begin. The peewits' cry for attention echoes round the lower fells, and the primrose and celandine tentatively push their way through the undergrowth to daub a bit of colour on the scene. Winter has left behind a stark brown landscape with no obvious sign that it is a time for new life. On days when the sun manages to dispel cloud cover, stretches of blue sky coax a richer tapestry out of the landscape, the air feels light and there is a hint of spring in the wind. The lambs lend their support to this image. They enliven the stark hills. They concoct races back and forth along the hedgerows, they jump imaginary hurdles, pretend ditches are treacherous gills, climb the stone walls, fight with the fencing and, when mother lies down, negotiate their way onto her back, knowing where the bed is best.

The other face of spring is bitter and drear. When the wind drops and gives some reprieve, the rain comes and the cycle of cold and wet seems interminable. The shepherd must devise shelter for the new-born lambs, as in hill country there may be little natural cover for them to escape exposure to the elements. Hay bales are stacked like children's bricks to make compartments in the barn for ewes with lamb problems; bits of fencing and the old gate are propped up outside to afford extra protection.

Exposure and starvation are the two main causes of loss during lambing. Wet coldness is the enemy of the new born lamb. If the birthcoat cannot get dried out, the lamb loses heat quickly, its vitality dwindles and unless warmth and shelter are provided, it will surely die. So the farm kitchen becomes a clinic and the weakly lambs are brought in to be kept by the fire or under a lamp, or even in the oven until they revive. As a last resort the shepherd may take the lamb between his legs, dangle it like a rabbit and using a syringe of dextrose solution, inject it with a slanting action directly in to the lamb's abdomen. When this works, the half-dead creature gets to its feet in moments.

Warmth is the first priority and food the second. The colostrum, which is the initial substance from the mother's teat, is vital to the lamb as it contains essential antibodies and nutrients. If the ewe has no milk, the shepherd may use colostrum from another ewe or from a cow to ensure that each lamb has protection. Once the shepherd is satisfied that the lamb has vigour to endure the conditions of exposure, and the strength and will to keep its tummy filled with milk, he can stand back and let the ewe get on with the job of mothering.

Shepherds and shepherdesses will go to considerable lengths to achieve this end. It is not unusual to have to go out at night to assist a ewe in difficulty, using the tractor lights to work by; there may also be the odd lamb that needs a nightly feed to hasten its revival and others that benefit from being topped up again last thing at night. A tired shepherdess, who also has to keep the larder well stocked, the washing under control, and children attended to, may falter with fatigue.

"I was goin' to feed t' lamb, wasn't I? It was near eleven o'clock and a very wet night, so I got mesel' all togged up. I pulled on this big woolly coat of my great aunt's 'cos all me other coats were soaked right through. I'd lamb milk in one hand and hay under t' arm and I set off for t' shippen. But it was rainin' hard and so slippy and dark, and at side of muck midden I lost me balance and went down in it. As I went, I hit me head on t' stone, and I ended up on me back looking at sky with feet stuck in t' air. Should 'ave seen me! The weight of me coat in midden kept me down and it started rainin' more, thunderin' and lightenin'. I called out but no-one could hear for t' wind. Couldn't move to get out. All that I could do was take me wellies off, slip me arms out of t' coat and roll o' er and o' er in midden to reach side. Me face, ears, hair, eyes, everything - I've never been so muck up. Imagine it, me sat in t' kitchen waitin' for t' water to heat. And t' lamb never got its feed."

But there are greater perils than even the muck midden. If a bacterial infection such as dysentery gets among the lambs, it can wipe out half a shepherd's crop. Indeed in any year there are bound to be a number of losses from ailments such as rattlebelly, watery mouth, pneumonia and swayback, and it is the shepherd's instinct, experience and hard work that will minimize these. Among the ewes the ailments can range from conditions such as twin lamb disease, caused by malnutrition in late pregnancy, to inflammation of the lamb-bed, bad udders, milk fever, malpresentations - a catalogue of possibilities. A sudden shock to the ewe such as a worrying dog or a fall may damage the foetus, and if it dies the mother may die too.

There are the freaks, but not many of them, and a two-headed or eight-legged lamb will probably have to be dismembered inside the ewe before it can be removed. If the lamb is lost, the shepherd's skill must be used to save the ewe. There are the rarer problems such as a 'twisted lamb-bed', when, however hard the ewe pushes, it cannot get the lamb out because the neck of the womb is twisted.

"I knew a shepherd once that'd put his hand in t' yow and t' other fella rolled yow o' er ya way and then rolled it back o' er t' other way and neck of womb would be straightened out. But mostly now you'd go to vet with that, and they just shove this injection in them – marvellous bloody stuff!"

The shepherd will strive to do all that he can himself without resorting to a vet's assistance but "ye can't abuse sheep for t' sake of a pund or two." So when a problem outfaces him, the ewe will be put in the Land-Rover and taken down to the surgery.

On some farms lambing will last for as long as eight weeks. By May there is a hint of green on the trees and hedgerows, and the shepherd can feel confident that winter is past. Snow still clings under the crevices, and where it has gone from the fell the grass is left brown and lacking lustre. As the shepherd takes his ewes and lambs back to the fell, he is aware that where previously the ground has been hard underfoot, there is at last some spring in it. The sun now has warmth and, as he gazes across the vast tracts of sky on the open fell, he feels uplifted and knows that a good day's work has been done.

CLIPPING AND DIPPING

There are lulls in activity in the farming year: June is one of them. It is a time for spring-cleaning the barns of winter muck, putting till on the fields, picking up stones and trash and trying to get a crop of hay to grow. The inside land is left alone now in the hope that sun and rain will quicken the growth of grass ready for an early cutting in July.

Although lambing is done, the shepherd will spend time watching his sheep when they first go back to the fell. A change of pasture can cause the ewes to drop down suddenly with a complaint known as 'staggers' or moss illness. Aware of this possibility the shepherd carries his injecting kit with him, and any sheep that is ailing will be given a dose of calcium, which usually cures it. There are also environmental ailments peculiar to certain fells such as 'canker' or 'saut', which is caused by eating certain herbage in bog areas. The lambs are affected by sores and "their eyes get fastened-to and lugs can drop off." Only a change of diet seems to arrest this, though shepherds will try their own treatments, such as keeping the lamb away from sunlight, or a dose of multivitamins.

When one ailment is cured, there always seems to be another waiting to strike and different years bring different problems. One year ticks will be bad; liver fluke and bracken poisoning the next.
"It's squirt, squirt, squirt, jab, jab, jab - and the buggers still die."

At first the lambs are protected through their mothers' milk but after three or four weeks, when they start grazing, they need dosing against worms. Following that, some time within the first three months of their life, they will have one vaccination against a multitude of ills ranging from pulpy kidney, tetanus and struck, to braxy and black disease. But treatment does not stop there: the cross-bred lambs need de-tailing, the male lambs need castrating, which turns them into 'wethers', and they must all be appropriately marked or 'smitted'.

According to the experience and preference of the shepherd, he will perform these acts in a variety of ways. Tails can either be cut off with a sharp knife, twisted off by hand, or 'rubber-ringed', in which case the tail is left to wither and drop off in due course. The methods of castration, or gelding, vary too. Some may prefer to use the technique of 'rubber ringing' the scrotum: this has

Swaledale Tup

the added advantage that when the older wethers need clipping, "ye get a clean sweep, for there's nowt there to get in t' way." Other shepherds have learned the skill of using nippers to crush the spermatic cords and associated blood vessels. One thing is certain: the fate of the wether is sealed, for castration makes for sweeter meat. Only a few of the best tup lambs will be kept for breeding purposes.

In Lakeland, where there is acre upon acre of open fell, the sheep belong to their own heaf. In other words they learn to know their bit of fell and stick to it, becoming acclimatized to its particular environmental characteristics. If the farm changes hands, the heaf sheep stay with the farm, and independent valuers reach agreement as to a fair price to pay for the stock. Much of the land in both the Pennines and Lakeland is referred to as common land but, contrary to popular belief, it is not public but privately owned. Shepherds use it 'in common with one another', each entitled to put on it such number of sheep as the 'stint' or 'gate' allotted to their farm allows.

As the land is unfenced and the sheep can wander freely, each farm has its own lug mark. Lug is derived from an old Norwegian word 'log', meaning a law, thus the shepherd stamps his lawful ownership mark on his animals by clipping a piece out of the sheep's ear. It is from this practice that an ear has come to be called a 'lug'. These marks are given different names depending on their design: fold-bitted, sneck-bitted, spoon-shanked, cropped and ritted, slit and so forth. They are carefully thought out so that the shepherd can fold the lug in a particular way and with "yan or mebbe two slags the mark can be clipped". And the lambs do not appear to object. The ownership of the sheep is then undisputed and each stray can be duly returned to the rightful farm. A 'white' lamb is one that is born on the fell and has no mark; a shepherd can only lay claim to it if it is accompanied by its mother who has the appropriate lug mark of his farm.

So shears need to be kept sharp at this time of year. When lug-clipping is done the tups and geld sheep are ready to be rid of their winter coats. The ewes that are still suckling their lambs must be left a while longer, for they do not grow wool the same as "yan that's doing nowt". Before sheep can be clipped there must be a 'rise' on the wool, and good feeding and fine weather help this process along. There is a saying:
"The new wool belongs to the sheep, and the old wool is the farmer's."

The coming of electricity changed the pattern of clipping time. When more shepherds were employed on the farms and the electric shears were not around to speed up the job, clipping time became a gathering together of neighbours who sat on their stools and hand-clipped their way through hundreds of fleeces, supping home-made lemonade to keep cool and eating pies to keep their arm strength up. Now the occasion varies from farm to farm depending on the number of sheep. Sometimes contractors, who specialize in shearing, are brought in to help if there are a few thousand sheep to work through, but otherwise father, mother and children will shear away until the job is done.

The younger generation will learn on the electric and with practice and perseverance considerable speed can be built up, though father, sticking to his hand-clippers will warn that "ye can take skin an' all unless ye're careful." It is tiring too: first hoisting the sheep up to turn it over, wedging it firmly between the legs, and then controlling it with only a limited length of cable. Backs and shoulders soon complain. The art is to cut the wool once and the sheep not at all. No sheep wants to end up teat-less!

The bane of clipping time, as with so many other farming activities, is wet weather, and also that it coincides with haytime. The fleeces cannot be stored damp because they rot, so the sheep must be reasonably dry before shearing; yet if they are brought into the barns for long, the sheep get so mucked up that the price of the fleece drops as the Wool Board objects to soiled wool. Once the fleece is off, it is rolled up from britch to neck and loosely bound with a length of neck wool. The fleeces are then ready to be collected and graded. They will be sold on to anything from the carpet or mattress trade to local weavers and knitters.

The shorn sheep gather together nervously. Having lost five pounds of wool from their backs they can take a while to readjust. Some shepherds will keep them on the inside fields for a week or so before returning them to the fell. This also gives a chance for their wool to grow a bit before they meet their next ordeal - the dipping tub.

Older sheep recognize the pervasive smell of disinfectant, know the routine, and defiantly hold back from being thrown into their tri-annual bath. Dips come in different shapes and sizes, but the idea is that sheep are gathered on one side in a holding pen, and then let through a swing gate one at a time to be thrown into the tub, where they are immersed for at least one minute. Some baths are designed like a canal, giving the sheep no option but to swim to the other end if they are to get out. This ensures that they stay in the mixture for a reasonable time.

By law it is required that the sheep are dipped against the external parasite, the scab-mite. But shepherds do not need convincing of the benefits of dipping. The nature of a sheep's fleece plays host to parasites such as lice, ticks, keds and blow-flies, the last of which can result in a mass of writhing maggots feasting on the sheep in an open wound. A sheep can be 'struck' to death by this pest. Ticks are another constant enemy and are worst on rough grassland. The dogs can get ticks too, but then dogs can also be dipped – indeed some vets encourage it. Most of them fall into the tub anyway now and again, as do the shepherds . . . as do some photographers!

The care, resourcefulness, and patience a shepherd shows as he moves amongst his flock cannot be evaluated. He has a respect for his sheep and their needs, and when they suffer they do not suffer alone. The shepherd will toil for many an hour to eradicate an ailment, or to minimize pain.
"Ye do not mind t' work, if it does t' job."

HAYTIME

As the sun and breeze sweep across the fell, lifting the grass heads and making patterns on the ground, the turmoil and dourness of the past winter and spring are fast dispelled. Summer haze on the distant hills melts land into sky, dividing lines vanish, and the fell gives one a feeling of uplift and space. Skylarks chatter, wheatears shift restlessly from rock to wall and back again and, by the stream, mauve-pink flowers of the native bird's eye primrose celebrate the Pennine summer. Land that seemed so bleak and desolate takes on new life.

The shepherd takes these contrasts in his stride. The same worn jacket and cap serve to protect him from the sun, cold or rain; he is consistent by nature. He knows the fell is not predictable and must never be taken for granted, but then the same could be said of the shepherd's lot; the fell appears to mirror it. Unexpected tragedy or a good sale, twin lambs or singles, wet ground or dry must all be accepted in good part so that a sense of non-attachment is achieved, enabling him to watch the different seasons and events, the ups and downs, with equanimity. The shepherd stands apart from those who fret and fuss. His one concern is for his sheep and they need steadfastness and regularity; they rely on him to give them that.

Down-dale the hedgerows echo the vibrancy of the upland hills. Roadside verges have become a lush background for cuckoo flower, clover, burnet and meadow-sweet, which manage to divert attention from the wide variety of orchids. The shepherd, aware of this intricate tapestry, remembers meanwhile that a few months from now the flowers will have gone, the frosts will have come, there will be little goodness in the grass and not much growth. His attention must turn from summer splendour to winter fodder, so he prepares to reap his harvest.

In the past decade attitudes and methods regarding haytime have changed. The unreliable summer weather on the northern hills means that the shepherd can never be sure of a few hot sunny days in succession when the grass is ready for cutting. The familiar story is that the grass is mown on a good forecast and a couple of days later, before baling up, an early morning shower undoes in one hour the weeks of care that have gone into preparing the crop. It may be a fortnight later before it is finally got in, and by then it has browned and lost much of its food value.

"Haytime's a hell of a job. There's nay bloody sun, and there isn't the men either. I reckon silage is a marvellous thing, old-fashioned though I may be."

The value of silage is that the grass can be got in when it is still damp, which is a godsend for the fell farmer. Nevertheless the farm has to be fit up for it. Silage is fermented grass, and to ensure that the curing process works efficiently it has to be either rolled up into big bales and stored in airtight plastic bags, or else dumped layer upon layer in a silage pit and squashed as flat as possible. To achieve this a tractor is driven back and forth on top of it, which can be a precarious business once it has reached a certain height; but if air gets into it, the silage goes mouldy and it is unfit for use. There also has to be some method of feeding it to the sheep during winter, as it can hardly be slung over the shoulder like a hay bale. In other words, silage needs method and machinery that may involve some reorganization at the hill farm.

Hay is still made when the time is right, and there is many a shepherd who prefers to feed his sheep with it: silage can be dangerous if it contains too much soil and it also has more bulk because of its moisture content. A good forecast and a ripe crop will persuade the farmer to hitch up the mower and start cutting as soon as morning milking is done. The following three days will then be spent retracing his tracks back and forth across the fields, first 'scaling' or whisking the cut grass with the hay-bob to help it to dry out, then rowing it up, and lastly baling it before extra hands are brought in to clear the field.

This is a time when family and friends pull together. Rakers are welcomed to tidy the scattered grasses and strong arms and backs are needed to stack the bales, heave them into a waggon and unload them into the barn. The more help there is on hand, the quicker the job is done, for there is always that nagging fear that rain clouds will move in. A week of haymaking and the body knows it. The neck is stiff from craning round to check that the hay-bob or baler is in line, hands are blistered where the baler twine cuts in, backs ache, eyes sting from the dust and there is an overwhelming sense of weariness as the last bale is got in close on midnight with everyone knowing that the whole process will begin again in a few hours time.

There are diversions though: haymaking is thirsty work and at regular intervals (if there is anyone left to organize it), a large wicker basket laden with squashes, cola, buns and biscuits will be taken out to the fields to keep up morale and energy. Children will skive off work to make hay-bale houses, burrowing inside them only to retreat itching before piling the bales and themselves onto the cart. Company, jokes and laughter temporarily ease the concern when the machinery breaks down and the heat begins to hurt. Headscarves and straw hats offer some protection, but basically making hay is a long, arduous, sweaty, dusty job without much reprieve.

Haytime does not just involve physical work: there is the decision making as well. One wrong move and three fields of good winter fodder could be lost. Hay or silage - which is it to be? Do we mow? Shall we bale?
"Sek a lot of argument. This year we were goin' to make big bale hay to save money on bags. The lad said, 'bag it, it'll rain tomorrow,' but then he doesn't have to pay for t' bags which cost a lot o' money when ye talleys it up. I wanted to wait - one more day and it would've made hay. Anyway he won and we bagged it and it rained the next day".

Hay or silage, good summer or wet one, winter fodder must be got in before it is too late. It is a time when little else is thought about. Once the first crop is taken, either animals will graze off the 'fog' or it will be left to grow a second crop which will be taken for silage later in the year.

Away from the sweat of haytime and the routine work on the farm, the tranquil aspect of the fells and dales in summer time cannot be ignored. White-washed cottages, cascading becks, cows ambling along lanes toward the milking shed, grazing sheep, all combine to paint a pastoral scene that serves to tempt off-comers to visit and stay a while. The shepherd and his family, always in need of extra revenue, will strive to accommodate them. If the farmhouse is big enough, comfortable beds and hearty breakfasts will be provided during the better weather months and the visitors are made welcome. There is a mutual exchange that benefits both parties. The fells, dales and fresh air are absorbed by the town-dweller, memories stored and taken away to be mulled over on winter days in homes and offices. And the visitor not only contributes to the farmhouse economy but also brings in different news and lends an inquisitive ear. Horizons are extended, ideas shared, and as the evening time is whiled away, many a lengthy yarn is spun.
"Hell we've had a good night - had a bloody good night. Fetched a bottle o' whisky for lile fella an' all. Nowt much wrong to be said about this summer."

SALES AND SHOWS

When the muted colours begin to hint at the drawing in of the year, the shepherd goes to the fell once more and brings in his flock to take the lambs off the ewes. Spaining or weaning time involves an 'in and out' gather. The sheep are again put through the sorting gates, but this time ewes go one way and lambs another. By now the ewes are in need of a break from their young before the breeding cycle begins once more, and they are turned straight back to the fell.

"It's the noise at spaining time that bothers me the most."
A vast pool of 'lile black Swardle faces' congregates in a large pen on the far side of the shedder and they loudly bleat their disapproval. They go in lambs and come out hoggs, a name they will keep until they are clipped for the first time the following July, when they become 'shearlings'. Shepherds manage spaining in different ways: some hoggs will be kept in overnight and by morning are so anxious for food that they soon forget their loss. Then they are taken to one of the lower fields well away from their mothers, for otherwise walls may be scaled and fences scrambled in attempts at reunion. As for the breeding ewes, they soon learn to recognize the cycle of events and accept it.

Herdwick Tups

In the following weeks the hoggs will be sorted into wethers, gimmers (yearling ewes), pure-breds, cross-breds, good stock and fair, and will be made ready for the big lamb sales that take place throughout the autumn. Wethers will go for fattening, some of the cross-breds will move further south to start their breeding cycle on lusher pastures, while the pure-breds will be kept as the shepherd's breeding stock. On the higher farms, few risk keeping their gimmer hoggs at home during their first winter; by November they will be sent to lowland farms, well away from the vicissitudes of the upland hills. They will return the following spring when, despite the months that have elapsed, it has been known for hoggs to re-find their mothers on the fell.

Milk cows, suckler calves, a few fat lambs and bed and breakfasters will earn farmers some income at different times of the year, but shepherds must wait until autumn for their proper financial reward. As the first of the main lamb sales approaches, and preparations are begun, there is an air of excitement around the farm. Sale time is when shepherds from many dales come together and there is an opportunity to discuss ideas, share news, comment on changing methods, and mull over the summer's good fortunes and losses. Sheep by the thousand pass under the hammer, the market pens are filled and crowds jostle.

It is a commercial and social gathering, a time when the year's hard work is assessed and valued in a few minutes in the ring, and the returns must last out the following year. Trade is never predictable.

The mule gimmer sales attract attention from many parts of Britain, for these lambs (Blue-faced Leicester tup and Swaledale cross) are hardy, easy to breed from and have attractively marked faces, an added bonus to lowland sheep farmers who like their sheep to look well on their land! Farmers from the south come north to buy them by the lorry load. The hill shepherd, aware of the extra value of mules and some other cross-breds, will take endless trouble to prettify them. If they look right and are a good size, the better they will sell. Their whiskers are trimmed to show off their heads, they are dipped in special formula to colour and curl their hair, faces are scrubbed to accentuate their markings, and a coloured ribbon is tied in their wool. Well-groomed, the sheep look very fine and, though it may be a "bloody carry-on", the shepherd will not complain if his pocket is better lined for his efforts.

On market day it is an early start for the shepherd who has some distance to travel and whose lot is early in the day. It is an occasion for a collar and tie and the smarter-looking cap, for presentation counts all round. On arrival at the mart, the sheep are herded into allotted pens where the shepherd will make the final adjustments. He will fluff up the fleece again to make it a good looking pelt, and once more apply the scrubbing brush and soapsuds with special added ingredient to heighten the distinction between white and dark wool. No amount of titivation is frowned upon.
"Why not bring t' hair dryer next time?"
"Aye, not a bad idea."

Elbows hooked over pens and legs propped on rails for support, critical eyes and earnest faces assess competition. The talk is of sheep and little else. There is a nervous banter as gossip is passed on between shepherds awaiting their turn. Finally the shepherd's lot is called and the sheep are coaxed along the far alley in readiness to enter the ring. The sing-song voice of the auctioneer concentrates the minds of buyers and sellers alike and, when it is time for the shepherd to make his entry, apprehensively he nudges his sheep around the sawdusted ring to show them off to their best advantage.
"Stand on, Lot 213, stand on now. A right good pen o' lambs from out top o' Dentdale. Scratched for orf, good stock and what'll anybody start out with?

Then a 50 to start, 50 for good sized lambs now. First bid at 50 from over there, bid's away at 50 - 51 - 52 - 3 - 4 - 5 - 6 . . . " One or two of the potential buyers standing at the side of the ring stop the shepherd to confirm details of the stock. He anxiously complies while the bidding goes on. "60 - 61 - 2, your bid stands at 62. Are you coming again? 62 bid, your bid stands at 62." And with a brisk tap of the stick, "sold to Mr . . . at 62 a piece. Next lot now . . . . "

Not much to grumble about at that price. Trade was good and the price is up on last year's. Relieved, the shepherd, now minus sheep, makes for the cashier's office to collect his cheque. The next stop is at the mart café where he settles down to a late breakfast of bacon, beans, and chips, served up by cheerful ladies who work day long to keep hearty appetites well satisfied.

Different marts, different weeks bring varied prices, so farmers will "sniff abut here an' there", watching trade and choosing carefully the right time and place to sell their stock. With a few hundred lambs to sell, the autumn weeks are spent to-ing and fro-ing between marts with trailer-loads of sheep. Debts are gradually cleared and the bank balance begins to fatten.

It is not just a time for selling but in Lakeland it is a time for showing sheep too. There is pride in good stock, and local shows and auction marts provide an opportunity for high standards to be aimed for and prizes to be won. In Lakeland, where there are a number of shows in different dales throughout the autumn, the Herdwick sheep are spruced up to show them off to best advantage. Custom decrees that the sheep are 'redded' and with good reason. "If a tip gets to ratchin' and gars wanderin' afore time, you can see it against t' grey stone if it's nicely redded up, and fetch it hyaam."
Red powder is sprinkled on the sheep's back and the shepherd as artist carefully brushes it in with a wetted scrubbing brush. There is style to the strokes and shepherds have their own ways of painting it on. Ewes as well as tips are now coloured up for showing.

Top tups, ewes, wether hoggs, gimmer hoggs, shearlings are all taken along to the show together with other farm produce. Cabbages and cakes, sheepdogs and crooks can all be judged and appropriately rewarded. Stock judging starts early and lasts the morning while stalls and side shows are set up for the afternoon's entertainment. Rain or shine folk will turn out and make the most of the day. The judges are selected locally and the greatest care and consideration are taken in choosing the winners, with due respect being paid to the entrants and their stock. Opinions inevitably vary regarding the best fleece, teeth, stance, walk and shape but prizes are generally spread around and, though some shepherds will end up better satisfied than others, there is always plenty to talk about. There is even a prize for the best sheep that has not won a prize!

When stock judging is done, consolation or celebration is sought in the beer tent, and it is time for the rest of the show to go on. There are still hounds, sheepdogs and the children's pets to be scrutinized and honoured and when the animals are done, people young and old begin racing each other up the fell and back to compete for the fastest time. Following the men it is the hounds who demonstrate their agility and speed in the characteristic sport of hound-trailing. Rags soaked in aniseed are trailed around the fell and the hounds follow the scent over scree, crag and dry stone walls until they reach the finishing line to a warm, noisy welcome from their owners. All this goes on to music playing and an endless succession of jokes from the compère, while children fill themselves with candyfloss and hot dogs. It is a day out in the year and shepherds, their families, locals and visitors alike make it into a veritable occasion.

When the show is done, empty pens, churned up fields, tents and tables are left to be cleared by the organizers. Come lambing time a shepherd may still find an odd coin in the field that has been dropped the previous year - a reminder of the festivity that is long gone.

Sports and shows over, the approach of winter now becomes a reality. It feels back-endish, and the winter dip needs doing before the cold weather starts to bite. It not only serves to protect the sheep against scabs and lice until the following spring, but also adds to the weather-resistant qualities of the fleece. This time the gather is easier, for there are no lambs on, and the ewes are familiar with the routine. The brackens, having turned a golden brown, appear to warm the fellside and give it a welcoming glow. Autumnal mists are beginning to creep in and the days are shortening fast.

As the shepherd, crook in hand, takes the familiar upward track to the tops, he scans the fellside for lone sheep, aware the light is softer now that the winter sun moves lower in the sky. The shepherd responds to the seasons, knowing there is no beginning nor end to his year. It is a cycle of change, an endless process that he latches himself on to. He pauses to button up his jacket for there is a chill in the air, then whistles at one of his dogs to fetch a couple of sheep down from the crag.

Yes, his work is his life and he would not change it. He belongs out here on the fell among his sheep . . . with the fireside a refuge at a day's end.

The shepherd is not alone in his affection for the fell. Over the years he has had to learn to share the land with all sorts of folk from ramblers and picnickers to climbers and adventure seekers. Between visitors and shepherds mutual respect is gradually being built up, but the process could be accelerated.

"We can strive to cure or prevent most ailments that our sheep get, but there's yan thing we canna' cure and there's yan thing we canna' prevent - and that's visitors and bloody dogs. They cause biggest losses of our sheep."
There is a passion in these words that goes deep. Offcomers, unaware that fell sheep move fast and have a gamey smell to them, believe their own dogs to be safe off the lead, but they can do untold damage.
"They chase sheep, hound them into becks, worry them 'til they're near to death, and then rive lumps of flesh out of 'em one after t' other. Dog owner doesn't even have the decency to finish job off - I'm left to shoot the sheep mesel".

Another problem is gates.
"Along comes a string o' folk - first opens gyat and by time last yan's through he thinks gyat's standin' open an' he leaves it. Two days tha can spend sortin' sheep out again."

Despite these problems, the majority of visitors are made welcome and their interest in the fells is appreciated. This interest is shared by the numerous organizations that also play active roles in the protection and care of the upland hill areas. The National Parks, National Trust, Nature Conservancy, other environmental groups, the Ramblers and the Ministry of Agriculture all have their own ideas as to how the fells and dales should best be served. The shepherd stands alone in the midst of them all.
"Jealousies creep in. Yan lot wants this and t' other wants that and the next lot don't want t' upset folk so they sit on t' fence like budgerigars. It's a real carry on I tell you, a road to nowhere. Yan day I had two dozen fellas in t' sitting room arguing abut me land."
Gradually some agreements are being reached between the different groups regarding conservation, access, fencing, grazing and so on, but it is a slow, laborious business. Understanding seems to be the key to effective change.

Fell farmers are in a unique position for there is little else they can do with their land except keep sheep. Changes in the nation's agricultural policy pose real

CONCLUSION

threats to their livelihoods as lowland farmers turn more to breeding fat lambs as an alternative to producing milk and cereals. The age-old role of the hill farmer may be seriously affected. Fell sheep will always be needed to maintain the resilience and hardiness of the nation's sheep stocks. It might also be helpful if we were more discriminating in our choice of meat, for the hill shepherd is capable of providing good lean lamb. A shearling that has been grazing on heather fell for two years produces a richer, sweeter meat: Herdwicks have earned themselves the title of 'King's mutton'.

It is a constant struggle for the hill farmer to be recognized and understood. Yet his whole life requires skill, dedication and endurance, and it is to be hoped these qualities will help him to overcome recent challenges, as he has done others in the past. It would be shameful if current changes were allowed to erode his livelihood.

As the shepherd views his life, he is not only concerned for the future but also nostalgic about the changing life in the valley. Shepherds' meets, village dances, whist drives, singing and making music at home are becoming pastimes of bygone days. As little as fifteen years ago, when there were still many farms without electricity, the family and friends would sit around in the evening making merry. Beer would be fetched in, the old fellows would sit on the settee playing dominoes by gaslight, the fiddle and accordion were brought out from under the stairs, and the kids would beat time on the biscuit tin with kindling sticks.
"Everybody's in sek a hurry today. They want more and more an' they've forgotten how to enjoy themselves. There's a time for everything, that's what I say. I could tell you . . . "

And there are many stories to be told.

THE PHOTOGRAPHS

"I allus think our year starts in November when we gather sheep in at tip time."

Barbondale

Tups are brought in at Malham

Ewes cross the face of Malham Cove

A shepherd scans the fellside for sheep in Upper Teesdale

The last of the ewes are brought down to Widdybank Farm

A Rough Fell tup is taken to his ewes near Sedbergh

"Ruddle is painted on his underside"

"The ewes are gathered to the tup to remind them all of their task" - Wharfedale

Back home to Starbotton

Farm stock and contents are sold off on the retirement of Jack Middleton

Feeding the hens

After an early snowfall near Askrigg

Heifers in Wensleydale

Betty Hartley's baking day

Broadmire Farm at Christmas

"He watches the weather as he watches over his sheep"

Spreading muck on the lower fields

Cows take a drink while their hill barn is mucked out

"There is little romance about winter on the fell" - Teesdale

Whernside

Kirkstone Pass

Wensleydale

Julie Burton leaves High Laning Farm

Age and infirmity: a veteran sheepdog, Tip, with Elizabeth who suffered a farm accident

Duncan Shuttleworth walling with the walkman for company

David Ellison puts gates on the new lambing shed

Sheep are counted to see how many are still on the fell - Cronkley Scar, Teesdale

Supplement is stepped up before lambing time

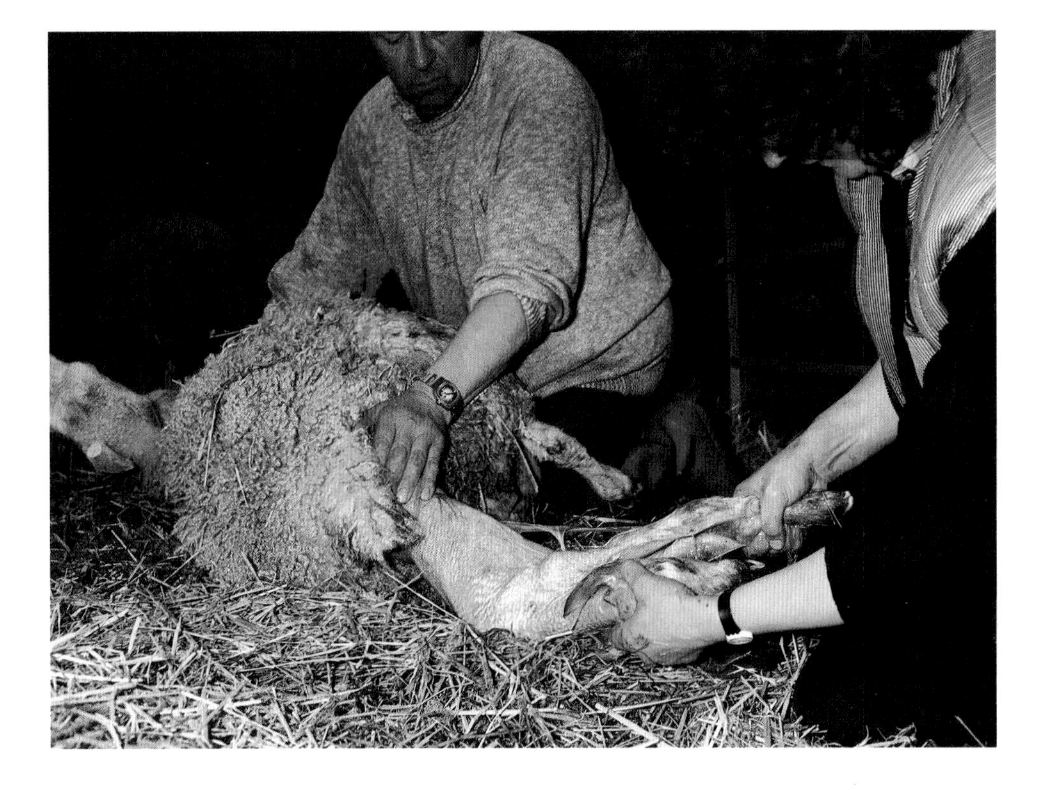

Lambing a Blue-faced Leicester

Twin tups Harold and Louis

The lambing shed in March

Harold and Louis at three days old

Harold needs a top-up from the goat

Gathering for lambing - Martindale

John Cockerill lambs a ewe in difficulty - Wharfedale

Marking newly born twins in Ribblesdale - 7am

At evening the older twins are moved to new pasture

The four-hourly feed

Protection from the cold and wet

The second week lambers are brought in over Kisdon Hill - Swaledale

Watching for difficulties in the lambing fields - Buttermere

A lamb's skin is needed for an orphan

An orphan is taken inside to be mothered on to a ewe that lost her lamb

A reluctant Herdwick mother needs tying up

Learning to accept

Martin Relph sets off on his third round of the day - Littletown, Newlands

A lamb needs help in learning to suck

Fresh blood indicates a fox is about

Week old lambs are gathered in for 'rubber-ringing' and marking - Newby Head

Off to the fell - Swaledale

Above Dentdale

Tan Hill Show

SLALOM
LAGER

Cows and calves are moved to new pasture in Wasdale - 6am

Herdwick sheep and lambs are driven down Langstrathdale

Nat Bland and Bob Cubby while sorting sheep

Topping and Tailing

'Scratching' against orf - a contagious disease causing sores

Grandfather, father, and son Mason gelding and marking - Deepdale

The Harrison family dosing and lug-marking - Eskdale

A sheep gather on Wandope fell near Buttermere

Planning a gather across the screes in Wasdale

"Shears need to be kept sharp at this time of year"

"Father uses hand clippers . . .

. . while son uses electric" - The Folders at Buttermere

Fleeces await collection in the barn

Haytime - Dentdale

The Thornborrow family haymaking near Keld - Swaledale

Loading up

Reg Charnley gathers sheep in for dipping

Big bag silage

Forking in the last of the grass

To market with the store lambs

Spectators at the sale

Jack and Ian Middleton at milking time

A cow with an awkward udder needs help

Jimmie Bentham

Len Middleton repairs a wall after a summer flood

"The shepherd scours the land from the tops"

"The sheep are driven down the uneven slopes" - Langdale

Gathering for spaining - Wharfedale

Ewes to the left and lambs to the right

The ewes are taken back to the fell - Langdale

A farm in Boredale changes hands and independent valuers assess the price to be paid for the stock

Preparation for the mule gimmer sale

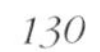

"No amount of titivation is frowned upon"

Mashams

Mules

The mule gimmer sale at Hawes Market

"A right good pen o' lambs from out top o' Dentdale"

"The next stop is the mart café"

The NFU at the Kendal Torchlight procession

Vet's work: an ailing house-cow and a newly bought calf need treatment

Margaret Taylor brings the hoggs in for dosing

Tups' pedicure

"Custom has it that Herdwicks are redded up"

Eskdale Show

"Rain . . .

. . or shine, folk will turn out and make the most of the day"

The judged

The judges

"There are still the sheepdogs and children's pets to be judged"

Tups are sold and hired out for the winter months at Keswick Fair

A few farms in the west suffer restrictions after Chernobyl

Troutbeck Auction Mart - an autumn lamb sale

A lame sheep is chivvied on to join the rest - Borrowdale Fells

"Heaped up and turned for home"

"They follow on in little lines"

The shepherd belongs on the fell . . .

NOTES ON THE PHOTOGRAPHY AND TEXT

The majority of the photographs were taken during 1988-9. Our aim was to accompany shepherds as much as possible as they went about their daily work, regardless of weather or time of day. The conditions were not always easy. On a steep hillside during a sheep gather it was often hard just to get shepherd, sheep and dogs in the picture, let alone have time to concern oneself about composition! We relied largely upon fast film - Fujichrome 400 - which allowed us to use high shutter speeds and so cope with fast action and overcast skies, as well as minimising camera shake. Wherever possible we used available light but inside dark barns we had to resort to flash, which we bounced off a home-made cardboard reflector fitted to our flashguns - this helped to soften the light.

Watendlath

On one very wet sheep gather we attempted to use our Mamiya 6x9 but despaired when faced with changing film every eight shots in the rain. After that we stuck to our Mamiya 645 cameras for most of the work. We then had the benefit of using roll film for better definition and finer grain, while having the greater flexibility of being able to shoot more frames on a film. We backed up with 35mm using Kodachrome 200 and 64 ISO.

The farmers were wonderful! We relied heavily on them informing us when they were gathering sheep, clipping ears, sharpening shears and so on. We were provided with nourishment, endless cups of tea and coffee, shelter, good company as well as a most welcome bed and breakfast at Watendlath. There, one night, we were regaled with stories and anecdotes until 4am by Dick Richardson and Bob Cubby, who both have made an invaluable contribution to the text. Otherwise information was gleaned from farmers throughout the year and advice was also given by Andrew Humphries at the Cumbria College of Agriculture. Our sincere thanks go out to all those who gave us their time and attention to help put this book together.